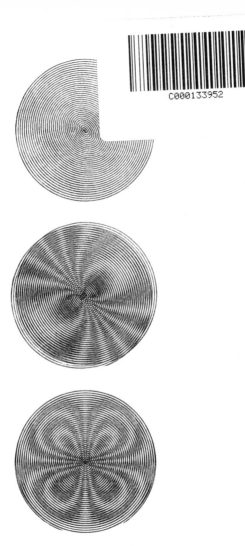

Unison (1:1): A spiral, a spiral drawn the same way over a spiral, and a spiral drawn the opposite way over a spiral.

This softcover edition © 2022 by Anthony Ashton
Published by Wooden Books LLC,
San Rafael, California

Hardcover edition first published in the US in 2003
by Bloomsbury Publishing USA, New York

Library of Congress Cataloging-in-Publication Data
Ashton, A.
Harmonograph

Library of Congress Cataloging-in-Publication
Data has been applied for

ISBN-10: 1-952178-04-5
ISBN-13: 978-1-952178-04-7

Designed and typeset in Glastonbury, UK

Printed in China on 100% FSC
approved sustainable papers by FSC
RR Donnelley Asia Printing Solutions Ltd.

WOODEN
BOOKS

HARMONOGRAPH

A VISUAL GUIDE TO THE MATHEMATICS OF MUSIC

Anthony Ashton

Dedicated to John, Antonia and Imogen.

Grateful acknowledgements to: *Harmonic Vibrations and Vibration Figures* by Joseph Goold, Charles E. Benham, Richard Kerr and L.R.Wilberforce, edited by Herbert C. Newton, Newton & Co. 1909; *Science and Music* by Sir James Jeans; Cambridge, 1937, *Sound* by John Tyndall, Appleton & Co., 1871, *Les Recreations Scientifiques* by Gaston Tissandier, Masson, 1881. The image on page 47 is taken from *Cymatics: A Study of Wave Phenomena and Vibration* by Hans Jenny, Macromedia Publishing, and used here by kind permission. I am particularly grateful to my grandson John Martineau for help with the intricacies of musical theory.

CONTENTS

Harmonic patterns from Sir Thomas Bazley's "Index to the Geometric Chuck" (1875),
showing concurrent & countercurrent phases with equal & inverted amplitudes.

INTRODUCTION

MANY OF THE DRAWINGS in this book were produced by a simple scientific instrument known as a harmonograph, an invention attributed to a Professor Blackburn in 1844. Towards the end of the nineteenth century there seems to have been a vogue for these instruments. Victorian gentlemen and ladies would attend *soirées* or *conversaziones*, gathering round the instruments and exclaiming in wonder as they watched the beautiful and mysterious drawings appear. A shop in London sold portable models that could be folded into a case and taken to a party. There may well be some of these instruments hidden in lofts throughout the country.

From the moment I first saw drawings of this kind I was hooked: not only because of their strange beauty, but because they seemed to have a meaning—a meaning which became clearer and deeper as I found out how to make and operate a harmonograph. The instrument draws pictures of musical harmonies, linking sight and sound.

However, before going any further I feel I should issue a health warning. If you too are tempted to follow this path, beware! It is both fascinating and time-consuming.

I should acknowledge my debt to a book called *Harmonic Vibrations*. It was coming across this book in a library soon after the end of the Second World War that introduced me to the harmonograph. Seeing that the book had been published by a firm of scientific instrument makers in Wigmore Street I went one day to see if they were still there. They were, though reduced merely to making and selling projectors.

I went into the shop and held up my library copy of the book for the elderly man behind the counter to see.

"Have you any copies of this book left?" I asked him.

He stared at me as though I was some sort of ghost, and shuffled away without a word, returning in a few minutes with a dusty, unbound copy of the book.

"That's marvelous," I said, "how much do you want for it?"

"Take it", he said, "it's our last copy, and we're closing down tomorrow."

So I have always felt that someday I must write this book.

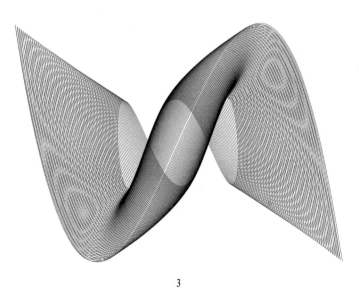

THE DISCOVERY OF HARMONY
on passing a blacksmith

To UNDERSTAND what the harmonograph does we need first to glance at the elements of musical theory.

Pythagoras, some 2,500 years ago, is credited with discovering that the pleasing experience of musical harmony comes when the ratio of the frequencies consists of simple numbers. A famous story tells of him walking past a blacksmith's shop. Hearing familiar harmonies in the ringing tones of the hammers on the anvil, he went in and was able to determine that it was the weights of the hammers which were responsible for the relative notes. A hammer weighing *half* as much as another sounded a note *twice* as high, an *octave* (2:1). A pair weighing 3:2 sounded beautiful, a *fifth* apart. Simple ratios made appealing sounds.

Although the myth is acoustically unsound, the picture opposite nevertheless shows the kinds of experiments the great philosopher supposedly went on to make (from Gafurio's *Theorica Musice*, 1492), as he found that all simple musical instruments work in much the same way, whether they are struck, plucked or blown.

Deeply impressed by this link between music and number, Pythagoras drew the metaphysical conclusion that all nature consists of harmony arising from number, precursor to the modern physicist's assumption that nature conforms to laws expressed in mathematical form. Looking at the picture you will see that in every example, hammers, bells, cups, weights or pipes, the same numbers appear: 16, 12, 9, 8, 6, and 4. These numbers can be paired in quite a few ways, all of them pleasant to the ear, and, as we shall see, also pleasant to the eye.

4

THE MONOCHORD OF CREATION
a singular string theory

There are seven octaves in the keyboard of a piano and nearly eleven in the total range of sound heard by the average person. The highest note of each octave has a frequency twice that of the first so the frequencies increase *exponentially*, on a *scale* beginning at 16 cycles per second (16 *Hertz*) with the lowest organ note and ending with about 20,000 per second. Below 16 Hz we experience *rhythm*. A range of ten octaves represents about a thousandfold increase in frequency ($2^{10} \approx 10^3$).

Robert Fludd's 17th century engraving of a monochord (*opposite*) shows the notes of the scale arranged in exponential fashion along a single string, with each assigned to planets or elements.

There is a hint here of what we can think of as the "great monochord" of the universe, also on a scale, this time stretching from a single quantum fluctuation at the bottom, to the observable universe at the top, passing through the various "octaves" of atom, molecule, quantities of solid, liquid and gaseous matter, creatures great and small, planets, stars and galaxies. Here too the scale is exponential, but usually measured in powers of ten, and covering a range of more than 10^{40}.

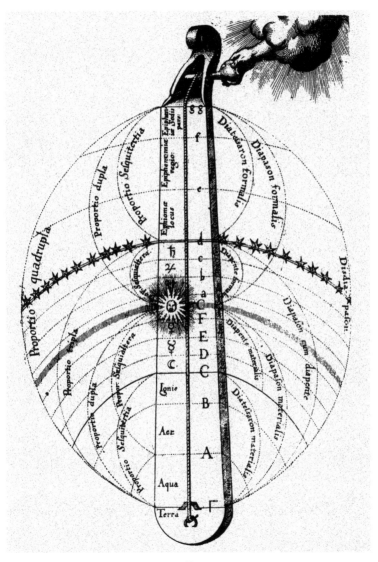

OVERTONES AND INTERVALS
harmonic ratios in and outside the octave

How are musical scales constructed? Listen very carefully as you pluck a string, and you will hear not only the main note, or *tonic*, but also a multitude of other harmonics, the *overtones*.

Overtones are any frequencies which are whole number multiples of the original frequency (*opposite top*). These harmonic resonances, 2:1, 3:1, 4:1 etc, affect not only strings and ringing hammers, but also columns of air and plates. Touching a string with a feather at the halfway or third point (*below*) encourages regularly spaced stationary points, called *nodes*, and an overtone can be produced by bowing the shorter side.

Musicians, however, need notes which are a little closer together than the overtone series, notes with *intervals* which harmonize within an octave. The lower diagram opposite shows the early overtone series on the left, with the corresponding intervals developing within the octave on the right, in order of increasing dissonance, or complexity.

"All discord harmony not understood" wrote Alexander Pope. The brain seems to grasp easily the relationships implicit in simple harmonies, an achievement bringing pleasure; but with increasing complexity it falters and then fails, and failure is always unpleasant. For most people enjoyment fades as discord increases. And, as we shall see, that is where the harmonograph drawings fade too.

tonic - open string

first overtone - up an octave

second overtone - a fifth above the octave

3rd overtone - up two octaves

Tonic, 1:1

The First Overtone
2:1 - octave

The First Interval
octave, 2:1

The Second Overtone
3:1 - fifth above the octave

The Second Interval
fifth, 3:2

The Third Overtone
4:1 - two octaves

The Third Interval
fourth, 4:3

The Fourth Overtone
**5:1 - two octaves
and a major third**

The Fourth Interval
major third, 5:4

The Fifth Overtone
**6:1 - two octaves
and a fifth**

The Fifth Interval
minor third, 6:5

9

TONES AND HALFTONES
the fourth and fifth get their names

Pythagoras' hammers hide a set of relationships dominated by octaves (2:1), fifths (3:2) and fourths (4:3). The fifth and fourth combine to make an octave (3:2 × 4:3 = 2:1), and the difference between them (3:2 ÷ 4:3) is called a *tone*, value 9:8.

A natural pattern quickly evolves, producing seven discrete *nodes* (or *notes*) from the starting tone (or *tonic*), separated by two *halftones* and five tones, like the Sun, Moon and five planets of the ancient world.

The interval of the fifth (3:2), the leap to the *dominant*, naturally divides into a major third and minor third (in Just tuning 3:2 = 5:4 × 6:5). The major third essentially consists of two tones, and the minor third of a tone and a halftone. The thirds can be placed major before minor (*to give the major scale shown in the third row opposite*), or in other ways.

Depending on your harmonic moves, or *melody*, different *tunings* appear, for example two perfect tones (9:8 × 9:8 = 81:64) are *not* in fact the perfect major third 5:4, but are slightly sharp as 81:80 (the *syntonic comma*, the Indian *shruti*, or *comma of Didymus*), more of which later.

Simple ratios, the octave and fifth, have given rise to a basic *scale*, a pattern of tones and halftones and, depending on where in the sequence you call home, seven *modes* are possible (*see page 52*).

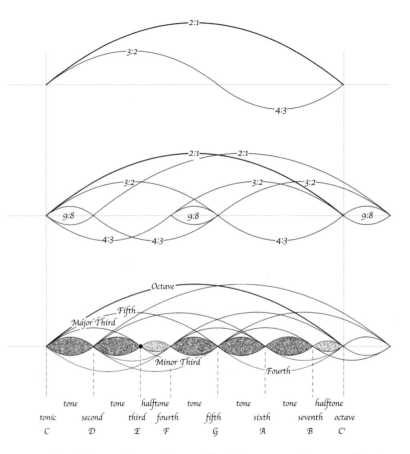

Above: The basic manifestation of the major scale. In Pythagorean tuning all tones are exactly 9:8, creating the leimma halftone of 256:243 between its major third (81:64) and the perfect fourth (4:3).

The sixth and the seventh are defined as successive perfect tones above the fifth. The picture is different in Just Intonation: here the major third is perfect at 5:4, which squeezes the second tone to 10:9 (a minor tone), leaving 16:15 as the Just halftone up to the fourth. The Just sixth is 5:3, a major third above the fourth, a minor tone above the fifth. The Just seventh (15:8) is a major tone above that, a major third above the fifth and a halftone below the octave.

ARRANGING THE HARMONIES
the power of silence

The simple ratios of the primary overtones and undertones can be plotted on an ancient grid known as a *lambdoma* (*opposite top, after Albert von Thimus*), after the greek letter λ. Some intervals have the same value (e.g., 8:4 = 6:3 = 4:2 = 2:1), and if lines are drawn through these it quickly becomes apparent that the identities converge on the silent and mysterious ratio 0:0, which is "outside the diagram".

A further device used by the Pythagoreans was the *Tetraktys*, a triangular arrangement of ten elements in four rows (1 + 2 + 3 + 4 = 10). The basic form is given opposite lower left, the first three rows producing the simple intervals. In a variation, the *lambdoma* (*opposite lower right*), numbers are doubled down the left side and tripled down the right, creating tones horizontally separated from their neighbours by perfect fifths. After the trinity (1, 2 and 3) notice the numbers produced, 4, 6, 8, 9, 12, and then look again at the picture on page 5.

Below we see a fuller range of monochord positions, with the overtones on the right, and intervals on the left (*see page 54-55*).

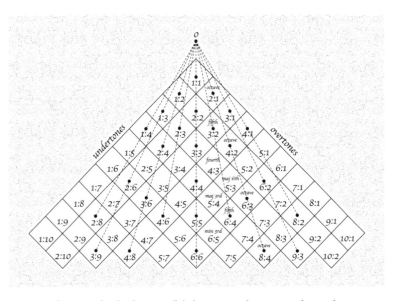

Pythagorean and medieval tunings, called 3-limit, recognized no true intervals except for ratios involving 1, 2 and 3. The lambdoma below right expresses this numerically as any element relates to any neighbour by ratios only involving 1, 2 and 3, so we can move around by octaves and fifths. Squares (4=2 2, 9=3²) and cubic volumes (8=2 3, 27=3³) also appear. Add further rows and the numbers for the Pythagorean scale soon appear: 1 9:8 64:81 4:3 3:2 27:16 16:9 2:1. This scale has four fifths and five fourths but no perfect thirds or sixths. These came later with Just Intonation and its perfect thirds (6:5:4), as polyphony and chords slowly took over from plainchant and drone.

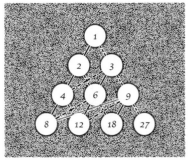

LISSAJOUS FIGURES
sound made shape

In the mid nineteenth century a French mathematician, Jules Lissajous, devised an experiment: He found that if a small mirror was placed at the tip of a tuning fork, and a light beam aimed at it, then the vibration could be thrown on to a dark screen. When the tuning fork was struck, a small vertical line was produced and if quickly cast sideways with another mirror it produced a sine-wave (*below*).

Lissajous wondered what would happen if instead of casting the wave sideways he were to place another tuning fork at right angles to the first to give the lateral motion. He found that tuning forks with relative frequencies in simple ratios produced beautiful shapes, now known as Lissajous figures.

On the screen (*opposite top*), we see the octave (2:1) as a figure of eight, and below it various *phases* of the major and minor third. These were some of the first fleeting pictures of harmony, which were doubtless familiar to Professor Blackburn when he devised the harmonograph.

Above: The Lissajous experiment. Small mirrors are placed on the tips of two tuning forks, one vibrating vertically, the other vibrating horizontally. A beam of light shone on the first mirror is given a horizontal wobble, and then a vertical wobble by the second tuning fork, combining to produce the Lissajous Figures.

Below: Small disharmonies cause the figure to work through its various phases, here 4:5 and 5:6.

THE PENDULUM
keeping time

A fundamental law of physics (in one formulation) states that left to itself any closed system will always change towards a state of equilibrium from which no further change is possible.

A pendulum is a good example. Pulled off center to start, it is in a state of extreme disequilibrium. Released, the momentum of its swing carries it through nearly to the same point on the other side. As it swings it loses energy in the form of heat from friction at the fulcrum and brushing against the air. Eventually the pendulum runs down, finally coming to rest in a state of equilibrium at the center of its swing.

Going back 500 years, Galileo, watching a swinging lamp in the cathedral of Pisa, realized the frequency of a pendulum's beat depends on its length: the longer the pendulum the lower the frequency. So the frequency can be varied at will by fixing the weight at different heights. Most importantly, as the pendulum runs down, the frequency stays the same.

Here, therefore, is a perfect way to represent a musical tone, slowed down by a factor of about a thousand to the level of human visual perception. For a simple harmonograph two pendulums are used to represent a harmony, one with the weight kept at its lowest point, while the weight on the other is moved to wherever it will produce the required ratio.

As we shall see, the harmonograph combines these two vibrations into a single drawing, just as two musical tones sounded together produce a single complex sound.

WORKING WITH PENDULUMS

The theoretical length of the variable pendulum that will produce each harmony can be calculated, for the frequency of a pendulum varies inversely with the square root of its length. This means that while the frequency doubles within the octave, the length of the pendulum is reduced by a factor of four.

Figures are given for a pendulum 80cm long, a convenient length for a harmonograph. These theoretical markers provide useful "sighting shots" for most of the harmonies. Note that the pendulum length is measured from the fulcrum to the center of the weight.

Interval Name	Approx. Note	Harmonic Ratio	Length (cm)	Freq. (s^1)
Octave	C'	2:1	20	66.0
Maj. 7th	B	15:8	22.8	61.9
Min. 7th	B♭	9:5	24.7	59.4
Maj. 6th	A	5:3	28.8	55
Min. 6th	G♯	8:5	31.2	52.8
5th	G	3:2	35.6	49.5
4th	F	4:3	45.0	44.7
Maj. 3rd	E	5:4	51.2	41.2
Min. 3rd	E♭	6:5	55.6	39.6
2nd	D	9:8	63.2	37.1
Halftone	C♯	16:15	70.3	35.2
Unison	C	1:1	80	33

When a pendulum is pulled back and then released, the weight tries to fall towards the center of the Earth, accelerating as it does so. As the pendulum runs down, the rate of acceleration, and so the speed of travel, is reduced, but in equal proportion to the distance of travel. The result is that the period (the time taken for two beats) or the number of periods in a given unit of time (the frequency) remains unchanged.

In the picture to the left the frequencies of beats x and y are the same. For the pendulum formula, see page 53.

Two Harmonographs
lateral and rotary

In the simplest version of the instrument two pendulums are suspended through holes in a table, swinging at right angles to each other. Projected above the table, the shaft of one pendulum carries a small platform with a piece of paper clipped to it, while the shaft of the other pendulum carries an arm with a pen.

As the pendulums swing the pen makes a drawing that is the result of their combined motion (*see left side opposite*). Both pendulums begin with the same length, further drawings can be obtained as one is then shortened by sliding the weight upwards and securing it with a clamp at various points. The harmonic ratios can be displayed in turn.

By using three pendulums however, two circular, or *rotary*, movements can be combined, with fascinating results (*see right side opposite*). Two of the pendulums swing at right angles as before, but are now both connected by arms to the pen, which in all rotary designs describes a simple circle.

Situated under the circling pen, the third and variable pendulum is mounted on gimbals, a device familiar to anyone who has had to use a compass or cooking stove at sea. Here it acts as a rotary bearing, enabling the pendulum carrying the table to swing in a second circle under the pen. As the pen is lowered the two circles are combined on the paper.

A further source of variation is also introduced here, for the two circular motions can swing in the same (concurrent) or opposite (countercurrent) directions, producing astonishing drawings with very different characteristics (*see appendices on pages 54–55*).

LATERAL

ROTARY

1:1

2:1

3:2

Above: Two harmonographs and some of the simple patterns they draw. On the left the simple lateral version and its patterns (open and closed phase); on the right the three-pendulum, rotary harmonograph and its drawings (concurrent and countercurrent). See too the twin-elliptic harmonograph illustrated in the lower left-hand corner of page 58.

SIMPLE UNISON 1:1
and the arrow of time

The simplest harmonograph drawing is produced when both pendulums are the same length and the table is stationary. With the pen held off the paper both pendulums are pulled back to their highest points. One is released, followed by the other when the first is at its mid-point. The pen is then lowered on to the paper to produce a circle developing into a single spiral.

If the two pendulums are released together then the result will be a straight diagonal line across the paper, the "closed" *phase* of the harmony, as opposed to the circular "open" phase. At intermediate phase points elliptical forms appear (*below*).

The running-down of harmonograph pendulums is an exact parallel to the fading of musical notes produced by plucked strings, and can also be thought of as graphically representing the "arrow of time" (*see opposite*), with the unchanging ratios of the frequencies representing the eternal character of natural law. The characteristics of the drawings result from the meeting of the running-down process with the "laws" represented by the various frequency ratios. We see that music, like the world, is formed from unchanging mathematical principles deployed in time, creating complexity, variety and beauty.

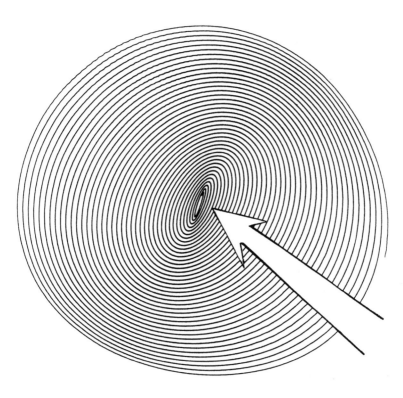

The inexorable direction of change, linked to the asymmetry of time (before-now-after), was vividly described by the scientist Arthur Eddington [1882–1944] as "the arrow of time." Throughout the process of continuing universal degradation, the dwindling stock of "useful" energy encounters a hierarchy of fixed physical laws conforming to mathematical formulae, and it is from the interaction of these unchanging laws with the arrow of time that comes a changing world of astonishing complexity, variety and beauty. The pendulum runs down from a state of disequilibrium to one of equilibrium, and the same is true, we are told, of the universe, the ultimate "closed system." From a state of extreme disequilibrium it plunged via the "Big Bang" towards its future ultimate state of utterly dark, frozen equilibrium. Between the beginning and the end there is a continual, cumulative transformation of "useful" energy, capable of forming temporary structures and causing events, into "useless" energy forever lost.

Near Unison
lateral phases and beat frequencies

A source of pleasing variety in harmonograph drawings comes from small departures from perfect harmonies. This seems to involve a principle widespread in nature as well as in the work of many artists. There is a particular charm in the near-miss.

An example from music suggests itself here. When two notes are sounded in near unison, the slight difference in their frequencies can often add richness or character to the sound. The two reeds producing a single note in a piano accordion have slightly different frequencies, the small departure from unison causing "beats", a subtle warbling or throbbing sound (*see page 53 for the beat frequency equation*).

Set the weights for unison and then shorten the variable pendulum slightly. Swing the pendulums in open phase, producing a circle turning into an increasingly narrow ellipse and then a line. If the pen is allowed to continue, the line will change into a widening ellipse, a circle, and a line again at right angles to the first. And so on. The instrument is working its way through the phases of unison shown on page 20.

If the variable pendulum is then further shortened in stages, a series of drawings like those opposite will be produced. The repetitive pattern represents "beats" with increasing frequency as the discrepancy between the notes widens. Eventually the series fades into a scribble that is a fair representation of discord, though even here there is a hint of some higher-number pattern.

For most people this fading of visual harmony occurs at about the same point as the audible harmonies fade.

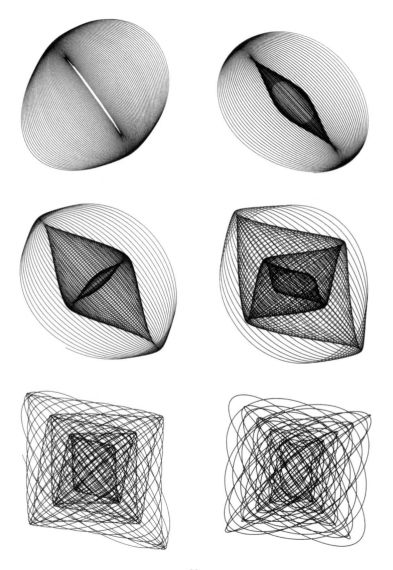

ROTARY UNISON 1:1

eggs and shells

At first this is disappointing: unison in contrary motion produces a straight line across the paper, like the closed phase of lateral unison. From concurrent motion there comes a mere dot that turns into a line struggling towards the center, pen and paper going round together.

However, changing to near-unison is richly rewarding. In contrary motion come a variety of beautiful, often shell-like, forms with fine cross hatchings. For best results lift the pen off the paper well before the pendulums reach equilibrium.

Surprisingly, from concurrent near-miss motion there come various spherical or egg-shaped forms. To produce an "egg" the pen should be lowered when it is dawdling at the center. It then spirals its way outwards, reaching a limit before returning as the pendulums run down. Because the lines toward the perimeter get closer together, the drawing appears three-dimensional.

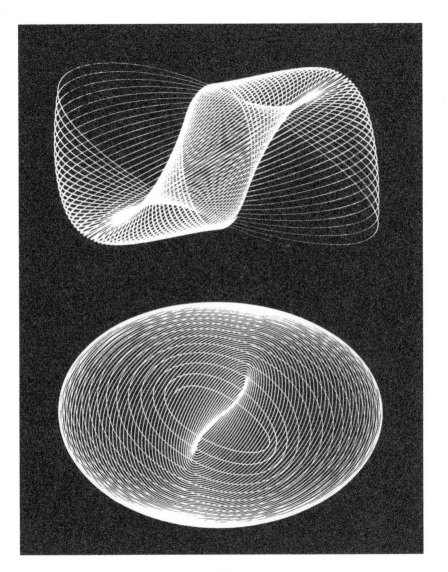

THE LATERAL OCTAVE 2:1
figures of eight and wings

After unison the next harmony to try is the octave. Here there is a technical difficulty, for the variable pendulum has to be very short, and because of the greater amount of friction involved runs down quickly. The trick is to add a weight to the top of the invariable pendulum, which slows it down (*see the picture on the title page*). The variable pendulum can then be longer.

Unfortunately this means that for the octave, and other ratios where one pendulum is going much faster than the other, the theoretical markers have to be ignored, and the right point found by trial and error.

With one pendulum beating twice as fast and at right angles to the other, the octave in open phase takes the form of a figure-of-eight (a coincidence), repeated in diminishing size as the pendulum runs down.

If both pendulums are released at the same time to produce the closed phase, the result is a cup-shaped line that develops into a beautiful winged form with fine cross-hatchings and interference patterns. Small adjustments produce striking variations.

The octave is the first overtone (*see page 8*).

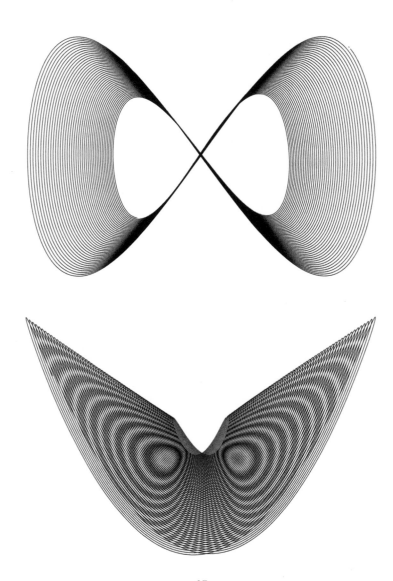

THE ROTARY OCTAVE 2:1
hearts and triangles

From rotary motion with a 2:1 ratio come some of the most beautiful of all harmonograph drawings: simple, graceful and often surprising. Remember, all that is happening here is that two circular motions, one almost exactly twice as fast as the other, are being added together.

Contrary motion produces a trefoil shape with many fine variations (*right hand images opposite*). Starting with a smaller size or *amplitude* in the faster rotation produces a triangle, or pyramid.

The octave in concurrent motion produces a heart-shaped form with a simple inner loop (*below left, and left hand images opposite*). Here there is a link with the ancient tradition of the music of the spheres, for this is the shape an observer on Uranus would ascribe to the movement of Neptune, or vice-versa. This is because the planets orbit the Sun concurrently, Uranus in 84 years and Neptune in 165, approximately performing an octave. The planet Mercury sings a perfect octave all by itself, as one of its days is two of its years.

Near-misses in the ratios of rotary drawings set the designs spinning (*lower images opposite*).

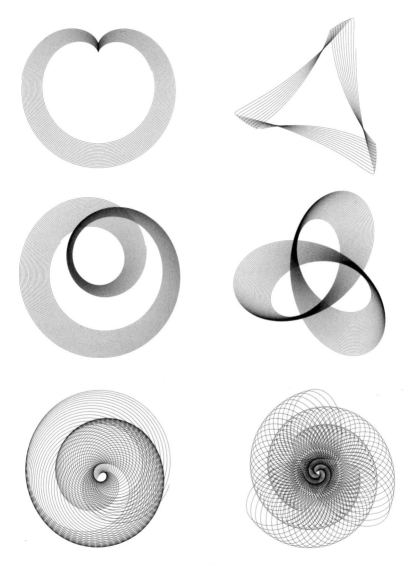

THE LATERAL FIFTH 3:2

and the second overtone 3:1

Next to be tried is the harmony of the fifth, intermediate between the simplicity of unison and octave and the more complex harmonies that follow.

It will be seen from the open phase drawing opposite that the fifth has three loops along the horizontal side and two along the vertical. The number of loops on each side gives the ratio, 3:2. Looking back at the octave, there are two loops to one, and with unison there is only one "loop", however you look at it. This is the general rule for all lateral harmonograph ratios, and if a harmony appears unexpectedly during experiments, it can usually be identified by counting the loops on two adjacent sides.

The fifth also appears as 3:1, the second overtone, a fifth above the octave (*see open and closed phase drawings of 3:1 on page 3*). Drawing ratios outside the octave may require a twin-elliptic harmonograph (*see page 58*). The phase-shifted pair below are stereographic; if you go cross-eyed they will jump into 3-D.

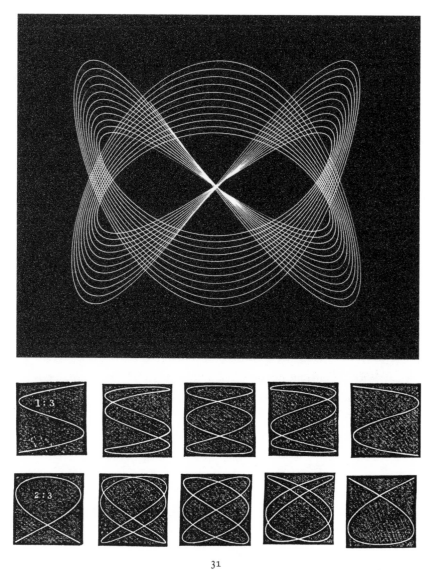

31

THE ROTARY FIFTH 3:2
encircled hearts and fives

The "loudness" of musical tones is represented on the harmonograph by *amplitude*, the relative sizes of the two circular motions. In rotary drawings this is much more important than phase, which simply orients the whole design on the page.

The third drawing below shows a rotary fifth in contrary motion where the higher-frequency, faster moving, pendulum has much the wider swing. In the "spiky" drawing to its right it is the other way round. At equal amplitude all lines pass through the center (*see table on page 55*).

The top four drawings opposite show rotary forms of 3:2, concurrent on the left, and countercurrent on the right. The second row shows the effect of a near-miss in the harmony, which makes the patterns spin.

The lower two images opposite are from the 1908 book *Harmonic Vibrations*. They show the second overtone, 3:1, a fifth above the octave (3:1 = 2:1 × 3:2), concurrent on the left, countercurrent on the right.

With concurrent pictures, the number of swirls in the middle is given by the difference between the two numbers of the ratio. So the concurrent patterns for the primary musical intervals 2:1, 3:2, 4:3, 5:4, and 5:6 all have a single heart at their center.

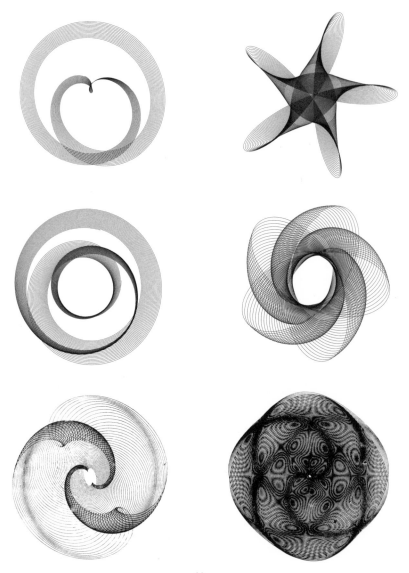

THE FOURTH 4:3
with thirds, sixths, and sevenths

By now it will be evident that each harmony displays its own distinct aesthetic character. Unison is simple and assertive. The octave introduces an emphatic flourish, and the fifth, while still fairly simple, has added elegance.

With the fourth the pattern becomes more complicated, though the design is still recognisable without counting the loops. The upper diagram opposite shows the fourth in open phase, the lower in closed phase. An increasing sophistication becomes apparent, and some of the closed phase and near-miss variants have a strange exotic quality.

Introducing perfect thirds increases the complexity. In Just Intonation the major third (5:4) is found below the fourth, the interval between them a halftone which works out as $4:3 \times 5:4 = 16:15$.

A fourth and a major third ($4:3 \times 5:4$) produce the *major sixth*, 5:3, a minor third (6:5) below the octave and a minor tone (10:9) above the fifth. Likewise, a fourth and a minor third ($4:3 \times 6:5$) create the *minor sixth* (8:5), a major third (5:4) below the octave and a halftone (16:15) above the fifth.

A fifth and a major third ($3:2 \times 5:4$) produce the *major seventh*, 15:8, while a fifth and a minor third ($3:2 \times 6:5$) give the *minor seventh*, 9:5.

These are the elements of the Just scale (*see page 50*).

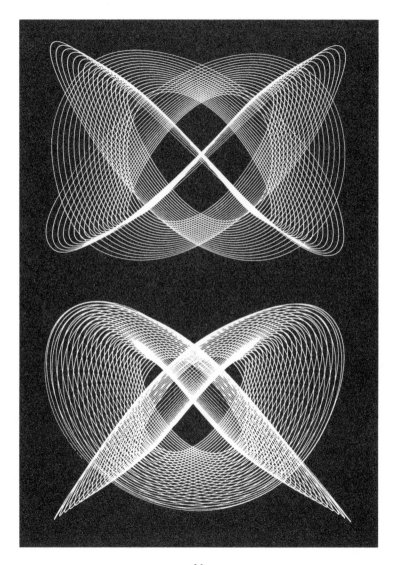

FURTHER HARMONICS
seven limit and higher number ratios

As the numbers in the ratios increase it becomes harder to distinguish the harmonies one from another at a glance: the loops have to be counted, and slight variations produce little of aesthetic value. A typical example, 7:5, is shown opposite top.

Rotary motion produces a series of increasingly complex drawings, influenced by relative frequency, amplitude and direction. In contrary motion the total number of loops equals the sum of the two numbers of the ratio. With concurrent motion the nodes turn inwards, and their number is equal to the difference between the two numbers of the ratio.

The contrary drawings below show a fourth (4:3), another fourth, a major sixth (5:3) and a major third (5:4). The lower pictures opposite, drawn over a hundred years ago, show unequal amplitude drawings of the perfect eleventh 8:3 (an octave and a fourth) and the ratio 7:3 which is found in seven-limit jazz tuning (not covered in this book).

Two octaves and a major third (4:1 × 5:4) equal 5:1, the fourth overtone, which differs from four fifths $(3:2)^4$ as our friend 80:81, the syntonic comma (*see page 10*). In *mean tone* tuning, popular during the Renaissance, this misfit was ironed out and the fifths were flattened very slightly, to $5^{1/4}$ or 1.4953, falling out of tune to please the thirds and sixths.

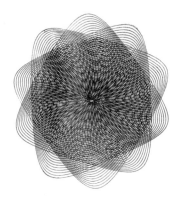

37

AMPLITUDE

circles, polygons, flowers and another circle

Much variation can be obtained from a rotary ratio by having unequal sizes in the two circular motions. Opposite we see two frequencies related by a major sixth (5:3). A lower frequency note begins to be influenced by, combines with, and is then more or less replaced by a higher frequency one. When the two notes are at equal volume the lines all pass through the center (*see pages 56–57*). Notice that the sequence is not symmetrical.

Below we see the first three overtones. For the spikiest shapes simply invert the amplitudes. For polygons, square them first.

If you have ever played with a "Spirograph", the harmony is determined by the cogging ratio, and the amplitude is adjusted by changing penholes on the wheel.

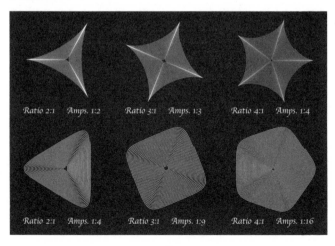

Ratio 2:1 Amps. 1:2   Ratio 3:1 Amps. 1:3   Ratio 4:1 Amps. 1:4

Ratio 2:1 Amps. 1:4   Ratio 3:1 Amps. 1:9   Ratio 4:1 Amps. 1:16

TUNING TROUBLES
the Pythagorean comma

Leaving the harmonograph drawings and returning to the principles of music, you may have noted that musical intervals do not always agree with one another. A famous example of this is the relationship between the octave and the perfect fifth (3:2).

In the central picture opposite, a note is sounded in the middle at 0, and moved up by perfect fifths to give the sequence C, G, D, A, E, etc. (*numbered opposite, each turn of the spiral representing a perfect octave*). After twelve fifths we have gone up seven octaves, but the picture shows that we have overshot the final octave slightly, and gone sharp. This is because $(\frac{3}{2})^{12} \approx 129.75$, whereas $(2)^7 = 128$. The difference is known as the Pythagorean comma, proportionally 1.013643, approximately 74:73.

If you kept on spiralling you would eventually discover, as the Chinese did long ago, that 53 perfect fifths (or *Lü*) almost exactly equal 31 octaves. The first five fifths produce the pattern of the black notes on a piano, the *pentatonic* scale (*see page 50*).

The smaller pictures opposite show repeated progressions of the major third (5:4), the minor third (6:5), the fourth (4:3), and the whole tone (9:8), all compared to an invariant octave.

It's strange. With all this harmonious interplay of numbers you would have expected the whole system to be a precisely coherent whole. It isn't. There are echoes here from the scientific view of a world formed by broken symmetry, subject to quantum uncertainty and (so far) defying a precise comprehensive "theory of everything". Is this why the "near miss" is so often more beautiful than perfection?

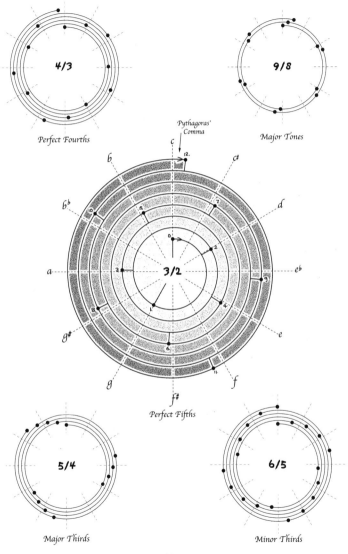

4/3

Perfect Fourths

9/8

Major Tones

Pythagoras'
Comma

3/2

Perfect Fifths

5/4

Major Thirds

6/5

Minor Thirds

41

EQUAL TEMPERAMENT
changing keys made easy

Although early tunings enabled many pure harmonics to be played, it was hard to move into other keys, one could only really change *mode (see page 52)*. Musicians often had to retune their instruments, or use extra notes reserved for specific scales (classical Indian tuning uses 22 notes).

In the sixteenth century a new tuning was developed which revolutionized Western music and which predominates today. The octave is divided into twelve fixed *equal* intervals, each *chromatic halftone* being 1.05946 times its neighbour ($2^{1/12}$, roughly 18:17).

The twelve equally spaced notes are arranged in a circle below. Six (flat) wholetones now make an octave, as do four (very flat) minor thirds, or three (sharp) major thirds. The Pythagorean comma vanishes, as do *all* perfect intervals except the octave—it's a clever fudge which allows us to change key easily. It is slightly "out of tune" and we hear it every day.

Triads are chords of three notes separated by thirds. Opposite top we see major and minor triads involving the note c, in the key of c. Use the mastergrid *(opposite, below)* to navigate the equal-tempered sea, and perceive any 3-4-5 triad in three distinct keys *(after Malcolm Stewart)*.

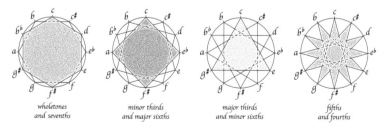

wholetones
and sevenths

minor thirds
and major sixths

major thirds
and minor sixths

fifths
and fourths

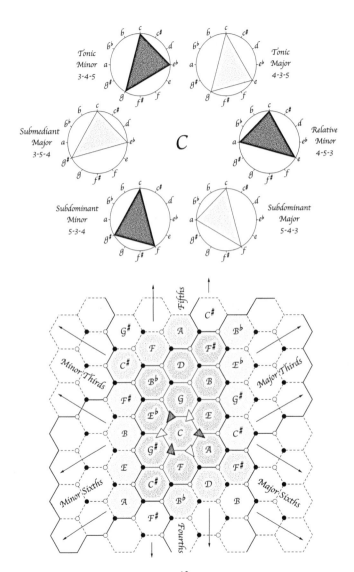

THE KALEIDOPHONE
squiggles from a vibrating rod

An interesting nineteenth century precursor to the harmonograph was the *kaleidophone*, invented by Sir Charles Wheatstone in 1827, who described it as a "Philosophical Toy". Like the harmonograph, it displayed optical figures of harmonics.

The simplest version of the device consists of a steel rod with one end firmly fixed into a heavy brass stand and the other fixed to a small silvered glass bead, so that when illuminated by a spotlight a bright spot of light is thrown upon a screen placed in front of it. Depending on how the kaleidophone is first struck, and then subsequently stroked with a violin bow, a surprising number of patterns can be produced (*a few are shown opposite*).

The kaleidophone does not behave like a string, as it is only fixed at one end. Like wind instruments, which are normally open at one end, the mathematics of its harmonics and overtones are slightly more complicated than the monochord or the harmonograph and the positions of the nodes are more variable (*the lower images opposite show some early overtones*).

Other versions of the kaleidophone used steel rods with square or oval cross-sections to give further patterns. Wheatstone used to refer to his invention as a "philosophical toy", and indeed, as we look at these patterns, it is easy to feel wonder at their simple beauty.

To make your own kaleidophone, try fixing a knitting needle into a vice and sticking a silver bead or cake decoration ball to the free end. Use or make a bright point light source.

45

CHLADNI PATTERNS
vibrating surfaces

So far we have only considered vibrating strings and other simple systems, but surfaces also can be made to vibrate, and they too can display harmonic or resonant patterns.

In 1787 Ernst Chladni found that if he scattered fine sand on to a square plate, and bowed or otherwise vibrated it, then certain notes, generally harmonics of each other, each gave rise to different patterns in the sand on the plate. As with the harmonograph, other disharmonic tones produced a chaotic mess. Sometimes he found that further patterns could be created by touching the side of the plate at harmonic divisions of its length (*shown below*). This created a stationary node (*like the feather on page 8*). Later work revealed that circular plates gave circular patterns, triangular plates triangular patterns and so on.

The six pictures opposite are from Hans Jenny's book *Cymatics*, one of the seminal texts on this subject. The vibration picture appears gradually, the sand finding its way to stationary parts of the plate as the volume steadily increases through the sequence.

RESONANCE PICTURES
and how to sing a daisy

A more complete set of Chladni figures is shown opposite, all two or fourfold because they were produced on a square plate.

Below, however, we see some circular pictures. They were photographed by Margaret Watts Hughes, a keen singer, in the 1880s on an ingenious device called an *eidophone*, which consisted of a hollow base with a membrane stretched across it and a tube attached to its base with a mouthpiece at the other end. As Mrs. Hughes sang down the tube, fine lycopodium powder scattered on the taut membrane suddenly came to life, bouncing away from some places and staying still at others, producing shapes which she likened to various flowers.

Yet again, we see recognisable forms and shapes appearing from simple resonance and harmony.

49

SOME FEATURED TUNINGS

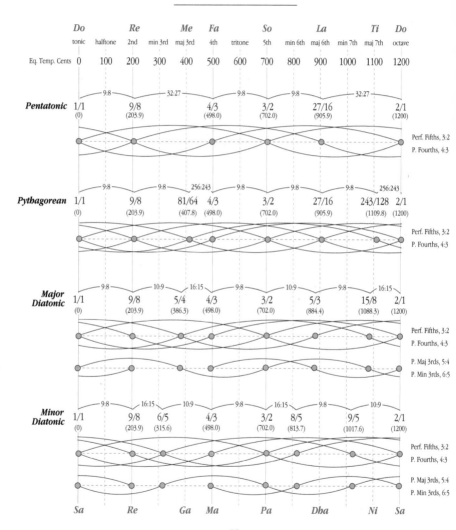

	Do		Re		Me	Fa		So		La		Ti	Do
	tonic	halftone	2nd	min 3rd	maj 3rd	4th	tritone	5th	min 6th	maj 6th	min 7th	maj 7th	octave
Eq. Temp. Cents	0	100	200	300	400	500	600	700	800	900	1000	1100	1200

Pentatonic

9:8 — 32:27 — 9:8 — 9:8 — 32:27

1/1 (0)　　9/8 (203.9)　　4/3 (498.0)　　3/2 (702.0)　　27/16 (905.9)　　2/1 (1200)

Perf. Fifths, 3:2
P. Fourths, 4:3

Pythagorean

9:8 — 9:8 — 256:243 — 9:8 — 9:8 — 9:8 — 256:243

1/1 (0)　　9/8 (203.9)　　81/64 (407.8)　4/3 (498.0)　　3/2 (702.0)　　27/16 (905.9)　　243/128 (1109.8)　2/1 (1200)

Perf. Fifths, 3:2
P. Fourths, 4:3

Major Diatonic

9:8 — 10:9 — 16:15 — 9:8 — 10:9 — 9:8 — 16:15

1/1 (0)　　9/8 (203.9)　5/4 (386.3)　4/3 (498.0)　　3/2 (702.0)　　5/3 (884.4)　　15/8 (1088.3)　2/1 (1200)

Perf. Fifths, 3:2
P. Fourths, 4:3

P. Maj 3rds, 5:4
P. Min 3rds, 6:5

Minor Diatonic

9:8 — 16:15 — 10:9 — 9:8 — 16:15 — 9:8 — 10:9

1/1 (0)　　9/8 (203.9)　6/5 (315.6)　4/3 (498.0)　　3/2 (702.0)　8/5 (813.7)　9/5 (1017.6)　　2/1 (1200)

Perf. Fifths, 3:2
P. Fourths, 4:3

P. Maj 3rds, 5:4
P. Min 3rds, 6:5

Sa	Re	Ga	Ma	Pa	Dha	Ni	Sa

SELECTED MUSICAL INTERVALS

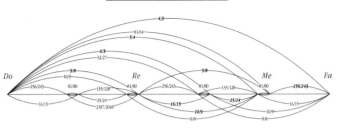

Interval	Cents	Name
1:1	0	*Unison*
32805:32768	2.0	*Schisma*
2048:2025	19.6	*Diaschisma*
81:80	21.5	*Syntonic Comma*
531441:524288	23.5	*Pythagorean Comma*
128:125	41.1	*Diesis*
25:24	70.7	*Minor Diatonic Halftone*
256:243	90.2	*Leimma, Pythag. Halftone*
135:128	92.2	*Major Chroma*
16:15	111.7	*Major Diatonic Halftone*
2187:2048	113.7	*Apotome*
27:25	133.2	*Large Leimma*
10:9	182.4	*Minor Tone*
9:8	203.9	*Major Tone*
8:7	231.2	*Septimal Tone*
7:6	266.9	*Septimal Minor Third*
32:27	294.1	*Pythag. Minor Third*

6:5	315.6	*Perfect Minor Third*
5:4	386.3	*Perfect Major Third*
81:64	407.8	*Pythag. Major Third*
4:3	498.0	*Perfect Fourth*
7:5	582.5	*Septimal Tritone*
45:32	590.2	*Diatonic Tritone*
729:512	611.7	*Pythag. Tritone*
3:2	702.0	*Perfect Fifth*
128:81	792.2	*Pythag. Minor Sixth*
8:5	813.7	*Diatonic Minor Sixth*
5:3	884.4	*Perfect Major Sixth*
27:16	905.9	*Pythag. Major Sixth*
7:4	968.8	*Harmonic Seventh*
16:9	996.1	*Pythag. Minor Seventh*
9:5	1,017.6	*Diatonic Minor Seventh*
15:8	1,088.3	*Diatonic Major Seventh*
243:128	1,109.8	*Pythag. Major Seventh*
2:1	1,200	*Octave*

Like the off-center division of the octave into fifths and fourths, sharps are not flats, giving rise to five more notes, making seventeen in all (found in middle-eastern tunings). More completely, we may think of the seven notes of the scale as moving across twelve "regions" of the octave, falling into the twenty-two positions of Indian tuning.

THE GREEK MODES

Modern Names	The Seven Modes of Antiquity		Ancient Greek Names
Ionian *Major*	do re mi fa so la ti do	c d e f g a b c 1 2 3 4 5 6 7 8	*Lydian*
Dorian	re mi fa so la ti do re	d e f g a b c d 1 2 3♭ 4 5 6 7♭ 8	*Phrygian*
Phrygian	mi fa so la ti do re mi	e f g a b c d e 1 2♭ 3♭ 4 5 6♭ 7♭ 8	*Dorian*
Lydian	fa so la ti do re mi fa	f g a b c d e f 1 2 3 4♯ 5 6 7 8	*Syntonolydian*
Myxolydian	so la ti do re mi fa so	g a b c d e f g 1 2 3 4 5 6 7♭ 8	*Ionian*
Aeolian *Natural Minor*	la ti do re mi fa so la	a b c d e f g a 1 2 3♭ 4 5 6♭ 7♭ 8	*Aeolian*
Locrian	ti do re mi fa so la ti	b c d e f g a b 1 2♭ 3♭ 4 5♭ 6♭ 7♭ 2	*Myxolydian*

The white notes on a piano approximate the seven notes of the seven modes of antiquity. Each mode, or scale, has its own pattern of tones and halftones, with two surviving as our major and natural minor scales. In reality various sizes of quartertones, halftones and tones were used. In addition, medieval transcription errors have left us with modern names which don't fit the ancient ones. Other scales include modal pentatonics, the harmonic minor with its minor 3rd and 6th, 1 2 3♭ 4 5 6♭ 7 8, and many others.

Harmonic Constants and Equations

The ratios and intervals in this book concern frequencies, normally expressed as cycles per second, or Hertz. Classical tuning sets C at 256 Hz. Modern tuning is higher, fixing A at 440 Hz. The period T of a wave is the reciprocal of its frequency f: $T = 1/f$.

The speed of sound in dry air is roughly $331.4 + 0.6T_c$ m/s, where T_c is the temperature in degrees celsius. Its value at room temperature, 20°c, is 343.4 m/s.

Gravitational acceleration on earth, g, is 9.807 m/s².

Frequency of a Pendulum.	$\dfrac{1}{2\pi} \sqrt{\dfrac{gravitational\ acceleration}{pendulum\ length}}$
Fundamental frequency of a tensioned string.	$\dfrac{1}{2 \times string\ length} \sqrt{\dfrac{string\ tension}{string\ mass \div string\ length}}$
Resonant frequency of a cavity with an opening.	$\dfrac{speed\ of\ sound}{2\pi} \sqrt{\dfrac{area\ of\ opening}{volume\ of\ cavity \times length\ of\ opening}}$
Fundamental frequency of an open pipe or cylinder.	$\dfrac{speed\ of\ sound}{2 \times length\ of\ cylinder}$

The beat frequency between f_1 and f_2 is the difference between them, $f_b = f_2 - f_1$.

The ratio $a:b$ converts to cents (where $a > b$): $(\log(a) - \log(b)) \times (1200 \div \log 2)$. To convert cents into degrees multiply by 0.3.

Clapping in front of a rise of steps produces a series of echoes with a perceived frequency equal to $v/2d$, where v is the speed of sound, and d is the depth of each step. Clapping in a small corridor width w produces a frequency v/w.

The *arithmetic* and *harmonic means* are central to Pythagorean number theory. The arithmetic mean of two frequencies separated by an octave produces the fifth between them (3:2), the harmonic mean producing the fourth (4:3).

6	:	8	::	9	:	12
$\mathcal{A}$	:	$\dfrac{2\mathcal{AB}}{\mathcal{A+B}}$	::	$\dfrac{\mathcal{A+B}}{2}$	:	$\mathcal{B}$
$\mathcal{A}$	:	Harmonic Mean	::	Arithmetic Mean	:	$\mathcal{B}$

Tables of Harmonograph Patterns

Overtone and simple ratio harmonics are shown below and opposite, arranged in order of increasing dissonance down the page. Open phase drawings display their ratio as the number of loops counted across and down. To find the ratio of a rotary drawing, draw both forms, concurrent (both circles in the same direction) and contrary (in opposite directions). Count the loops in each, add the two numbers together and divide the total by two. This gives the larger ratio number. Subtract this from the contrary total to give the lower ratio number. Alternatively, note that rotary figures for the ratio *a:b* will have *b-a* loops and *b* crossing places when both circles are concurrent, and *a+b* loops when they are contrary. The designs shown here were all made with equal amplitudes.

	lateral		intervals	rotary	
open phase	closed phase			con-current	counter-current
			1:1 unison		
			2:1 octave		
			3:2 fifth		
			4:3 fourth		
			5:3 major sixth		
			5:4 major third		
			6:5 minor third		
			8:5 minor sixth		
			9:8 wholetone (second)		

BUILDING A HARMONOGRAPH

ANYONE SERIOUSLY INTERESTED in making a harmonograph should consider going straight for the three-pendulum model.

The table must be highly rigid and firm on the floor, otherwise the movements of the weights will be distorted. I suggest it should be about 90 cm above the floor, 60 x 30 cm for two pendulums, 60 x 60 cm for three, and some 2 cm thick with an apron all round, about 8 cm deep.

Legs should be about 6 cm square, slightly splayed outwards and pointed at the bottom. One way of achieving the splay is to fix wood or metal brackets in the corners under the table each side of the diagonals and bolt the legs between them. After adjusting the legs to give the correct splay they can then be fixed in position with screws through the apron (which can be much deeper than shown).

To save space, slice off the table as along the dotted line. Three legs are not quite so stable, but work fairly well.

The platform carrying the paper should be light and rigid, fixed to the pendulum with a countersunk screw. A size of 22 x 16 cm will conveniently take half an A4 sheet of paper secured by a rubber band or small clip.

All sizes suggested are maxima, but a scaled down version will still work if it is carefully made.

If you are tempted to make a harmonograph, start with the weights, for the instrument will only be satisfactory if these are really heavy and yet easy to adjust. It is a good idea to make about ten of, say, two kilos each, so the loadings can be varied. They should be about 8cm in diameter, with a central hole, or better with a slot for easier handling. Either cast them yourself from lead (using old round cheese boxes) or ready-mixed cement or have them made by a metal smallware dealer or friendly plumber.

The shafts should be made from wood dowel, about 1.5 cm in diameter (metal rods are liable to bend, distorting the drawings), marked off in cm.

Building a Harmonograph

Clamps can be obtained from suppliers of laboratory equipment. For some of the drawings top weights are needed, held in place by clamps. Clamps can also be added to pendulum tops for "fine tuning" with one or more metal washers added.

The simpler kind of bearing consists of brass strips bolted into a slot in the pendulum and filed to sharp edges to rest in grooves each side.

In a bearing involving less friction the pendulum is encased at the fulcrum in a horizontal block of hardwood with vertical bolts each side filed to sharp points and resting in dints in metal plates. If drilling the large hole in the block is too difficult, it can be made in two halves, each hollowed out to take the shaft and bolted together.

Rotary motion needs gimbals. Here the grooves for the pendulum are filed in the upper side of a ring (e.g. a key-ring) while the under side has grooves at rightangles to the upper ones. The lower grooves fit on two projecting sharp edges (brass strips), each enclosed between two pieces of wood fixed to the table. With the alternative bearing a large flat washer should be used with depressions to take the sharp points.

Pen arms should be as light as possible to minimise "top-hamper" They are easily made from balsa-wood strips sold by model-making shops, using balsa cement and scotch tape.

For two pendulums the arm can be fastened to the shaft with pinched-off needles, and the pen jammed into a hole at the other end. For three pendulums the side pieces on the arm should enclose its shaft firmly but not too tightly and be held gently with a thin rubber band. One of the arms holds the pen, while the other is held by protruding needles pushed in backwards and secured (gently) at both ends by the rubber bands.

There may well be better, more sophisticated ways of doing all this. All suggestions welcomed.

An additional fitting is needed to lock a rotary pendulum so that the instrument can be used with

Building a Harmonograph

just the two single-axis pendulums. This can be done by mounting two brackets on the table near the rotary pendulum with holes to take a long horizontal bolt (slightly to one side) to which the shaft can be clamped.

Pens should be fine, light and free-flowing. Most stationers and shops selling draughtsmen's and artists' materials offer a variety (avoid ball or thick fibre pens). For best results use shiny "art" or "imitation art" paper, ordinary copy-paper for preliminary experiments.

If the pen is left on the paper to the end there is usually an unsightly blob. To avoid this, mount a short pillar on the table with an adjustable lever carrying a piece of thin dowel placed under the pen arm. By raising the dowel gently the pen is lifted off the paper without jogging it. This device should also be used before the pen is lowered to the paper. By watching the pen you can see what pattern is being made, and nudge it one way or the other by pressure on the pendulums.

For ratios outside the octave, such as 4:1, you may need to try another harmonograph such as Goold's twin-elliptic pendulum (*see left*).